Leiliane Cristina Lopes

Rainfall impacts in the urban area of Rio Claro - SP

Leiliane Cristina Lopes

Rainfall impacts in the urban area of Rio Claro - SP

Analysis and mapping of occurrences from 2005 to 2010

ScienciaScripts

Imprint

Cover image: www.ingimage.com

This book is a translation from the original published under ISBN 978-3-330-75810-0.

Publisher:
Sciencia Scripts
is a trademark of
Dodo Books Indian Ocean Ltd. and OmniScriptum S.R.L publishing group

120 High Road, East Finchley, London, N2 9ED, United Kingdom
Str. Armeneasca 28/1, office 1, Chisinau MD-2012, Republic of Moldova, Europe
Printed at: see last page
ISBN: 978-620-8-29560-8

SUMMARY

DEDICATORY

I dedicate this work to my parents and my brother for their constant support and affection, without which this achievement would not have been possible. I would also like to dedicate it to my friends for all the help they have given me, which has been fundamental to the completion of this work.

ACKNOWLEDGMENTS

To Prof. Dr. Anderson Luis Hebling Christofoletti, for his guidance, support and attention throughout this work.

To Carlos Augusto da Costa Prochnow, technician at the climatology laboratory of the Department of Geography at Unesp, Rìo Claro, for his indispensable help in collecting data and mapping it.

To Lieutenant Kleber Moura de Oliveira, Commander of the Rio Claro Fire Station, for his kindness in allowing me to collect the occurrences in the archive and to the members of the Fire Department for the attention and promptness with which they attended to me during the collections.

To my friend Bianca Falsarella, a colleague from my scientific initiation program, for her support, patience and fundamental help, which made this work so much easier.

To all the friends who, directly or indirectly, helped to make this work a reality by showing their support and patience.

SUMMARY

Urbanization is a marked process in our society, present since around 3,500 BC, in Mesopotamia, the area of present-day Iraq. Since the Industrial Revolution, cities have seen a significant increase in growth and have concentrated more and more labor and capital. Brazil, despite its late industrialization, also had a high population increase in its cities, especially in the 18th century, which were not provided with any kind of planning aimed at their order and the implementation of essential urban devices. As a result, Brazilian cities began to suffer from countless structural and environmental problems. Associated with inadequate infrastructure, changes in the urban atmosphere, which are related to increased heat, air pollution and, in particular, increased rainfall and storm frequency, cause various types of impact on cities. Among the most worrying are those caused by heavy rainfall, which cause material and human damage. The city of Rio Claro (SP) can be considered an example of this reality. The aim of this study was to analyze the impacts of rainfall on the population in the urban area between 2005 and 2010, by mapping the occurrences and analyzing the atmospheric systems involved. The information on occurrences was collected from the Fire Department and the press, while the synoptic charts were obtained from the INPE library in Cachoeira Paulista (SP) and from the Brazilian Navy website. The data showed that of the 247 occurrences over the six years, 233 happened in the spring-summer period. In addition, after mapping, it was possible to verify the locations considered to be critical points in terms of the frequency of occurrences, which were: Visconde do Rio Claro Avenue and the Central Zone, both with a total of 23 occurrences, the Jardim Inocoop neighborhood, with 20, Tancredo Neves Avenue, totaling 12, and Vila Paulista, with 10 occurrences. Using the graphs, it was possible to see that the most recurrent atmospheric system during the period studied was that related to convective precipitation; however, analyzing the years separately, other types of events predominated.

Keywords: Urban Climate, Geographic Climatology, Rainfall Impacts

I - INTRODUCTION

Urbanization is a very old phenomenon, which originated in Mesopotamia, the area of present-day Iraq, around 3,500 B.C. Sposito states that,

"Mesopotamia was (...) the center of the diffusion of the urban fact to Ancient Egypt (Memphis and Thebes), the Indus Valley (Mohenjo-Daro), the Eastern Mediterranean and inland China (Beijing and Hang-Chu)". (1996, p. 19)

Cities have spread and become increasingly important throughout history and have come to dominate gradually larger areas. After the Industrial Revolution, in the middle of the 18th century, the process of urbanization intensified considerably due to the urban nature of industrial production. Cities then became their territorial base and began to grow rapidly, as they were where capital and the workforce were concentrated. Sposito concludes that,

"The beginning of industrialization, understood here as a feature of contemporary society, as the main economic activity and the main way in which society appropriated nature and transformed it, profoundly marked and revolutionized the very process of urbanization". (1996, p. 48)

In the case of Brazil, the formation of cities began with Portuguese colonization, first on the coast and then expanding inland. Santos (1993, p. 2122) argues that it was from the 18th century onwards that urbanization developed in the country, although it was only consolidated in the second half of the 20th century. In the 1950s, with the implementation of Juscelino Kubistchek's development plan, industrial activity was emphasized. Then, in the 1970s and 1980s, as Santos (1993, p. 33) states, "the numerical growth of the urban population was already greater than that of the total population. The process of urbanization accelerated and reached a new level, which was consolidated in the following decade".

This process of accelerated and disordered urbanization has led to a number of consequences and transformations in the structure of cities, which have not been provided with any kind of planning aimed at implementing essential urban devices for the population.

The rapid and chaotic growth of cities has led to a drop in the quality of life of the population, which has had to deal with various problems related to basic infrastructure, such as water, sewage, electricity and street paving. It can also be said that it was in the cities that several environmental problems were concentrated, including river pollution, deforestation and, with regard to the atmosphere, air pollution, temperature changes and increased rainfall. These changes in the weather have led to significant alterations in the climate of cities, which have come to have a climate of their own.

The climate of cities is characterized by the imbalance between atmospheric elements and the earth's surface caused by the countless transformations imposed on the landscape by human action. The increase in rainfall in cities is one of the characteristic alterations to the urban atmosphere that cause worrying impacts and immense inconvenience to the population.

Among the many impacts caused by precipitation, the most common are: damage to street and avenue pavements, landslides, landslides and floods. Floods, in turn, cause traffic congestion and consequent travel delays, drowning and a large number of homeless people.

The material and human damage is extremely severe, which leads to the disorganization of cities and, consequently, to a reduction in the quality of life of their population. This highlights the vulnerability of cities due to inadequate infrastructure, the lack of integrated planning and efficient management aimed at solving urban problems.

Thus, the importance of urban climate studies, with a focus on rainfall, is highlighted by various authors, especially in relation to countries with tropical and subtropical climates, where rainfall is the element of the climate that causes the most rapid transformations in the landscape, especially in the summer, when there are episodes of intense rainfall.

Among them, Monteiro and Mendonça (2011, p. 55) state that there is hardly a year, especially in countries like Brazil, when no city is hit by floods, highlighting, above all, the inefficiency and inability of public authorities to combat the problems caused by intense rainfall. He therefore points out that

"The complexity of the issue and its national importance call for greater consideration of both the genesis of the phenomenon and its urban impact on defense mechanisms" (MONTEIRO; MENDONÇA, 2011, p. 55).

Ayoade (1986, p. 320), on the other hand, points out the growing need for investment in scientific studies on the atmospheric environment in developing tropical countries. Efficient climate planning that takes into account atmospheric characteristics and processes and their relationship or interaction with human biological and socio-economic activities is extremely important.

It is clear that the impacts of rainfall are increasingly manifesting themselves in Brazilian cities, pointing to the need for in-depth analysis of the problems most commonly related to heavy rainfall. These studies are essential in order to facilitate the actions of public authorities with regard to the planning and organization of urban space.

The city of Rio Claro (SP), due to the presence of built-up areas in its center, the various neighborhoods with industrial establishments and the continuous movement of motor vehicles on the main avenues, has significant changes in its urban atmosphere, which cause a greater intensity and frequency of rainfall and, consequently, various types of impacts on the population.

In short, it is extremely pertinent to study the location and different types of rainfall impacts in the urban area of Rio Claro (SP), making it possible to know the most problematic points in the city, in order to facilitate the taking of measures by the public authorities to solve the problems encountered.

II - THEORETICAL FRAMEWORK

The climate has long aroused human interest, as it is constantly influenced in its many activities by atmospheric conditions, which in turn can be influenced by human activities. According to Drew (1994, p. 73), "man first altered the local action of the atmosphere, and therefore the climate, 7 or 9 thousand years ago, by changing the face of the earth with the felling of forests, sowing and watering".

The first big leap in the development of atmospheric science occurred during the Renaissance, with the invention of the thermometer by Galileo in 1593 and the discovery of the principle of the mercury barometer by Torricelli in 1643. The invention of the telegraph in 1832, in turn, helped meteorological data to be transmitted more efficiently, making it easier to analyze at more widely located stations.

Initially, Climatology was only descriptive and generalizing, using average maps and graphs that showed diurnal and seasonal variations, as well as spatial differences in the values of the various climatic elements. This type of approach was characterized by what was called Traditional Climatology.

The purely descriptive approach of traditional climatology studies has numerous shortcomings, mainly related to the fact that it considers the atmosphere to be static rather than dynamic. Therefore, there is a need to understand atmospheric phenomena, taking into account their constant turbulence.

Sorre (apud MONTEIRO, 1976, p. 22), in defining climate as "the series of atmospheric states above a place in their usual succession", adopts a dynamic conception, which considers the set of climatic elements acting in combination. This gave rise to the notion of "types of weather", which consists of atmospheric conditions that recur with a certain frequency. This concept characterizes the so-called Geographical Climatology, which inaugurated a greater concern with studies on more detailed scales, from micro-, topo- and even meso-climatic, as well as understanding the genesis of climatic phenomena and their respective repercussions in space.

Based on Sorre's conception of climate, Monteiro (1976) proposes a methodology based on the notion of rhythm to investigate climate and concludes that in order to understand the variations in climatic elements, it is necessary to associate them with regional atmospheric circulation and correlate them with other geographical factors. It can therefore be said that Monteiro's (1976) work guided Brazilian Geographical Climatology, providing important concepts, among them, as Zavattini (2004) points out, the adoption of the Sorrean concept and the advent of the new paradigm related to rhythm, the focus on the dynamics of air masses and types of weather, the need to analyze climate on different scales and the importance of atmospheric dynamics.

Geographical Climatology is directly related to climate change, and its function is to understand the various alterations that have been taking place in the urban atmosphere.

Monteiro (apud MONTEIRO; MENDONÇA, 2011, p. 19) conceives the urban climate as "a system that encompasses the climate of a given terrestrial space and its urbanization". He therefore proposed a specific methodology for studies related to the climate of cities, the Urban Climate System (U.C.S.). According to the author,

"The S.C.U. imports energy through its environment, it is home to a succession of events that articulate differences in states, changes and internal transformations, to the point of generating products that are incorporated into the core and/or exported to the environment, configuring itself as a whole of complex organization that can be classified in the category of open systems." (MONTEIRO; MENDONÇA, 2011, p. 20)

It can be said that the S.C.U. aims to understand the particular climatic organization of the city, admitting the correlation of its different compartments (ecological, morphological or urban functional). In this way, it is necessary to integrate its different elements in order to characterize them with attributes whose relationships depend on the performance and functional organization of the system.

Due to the fact that variables such as heat islands, atmospheric pollution and flooding of the urban space stand out as the main consequences of modifications to the urban atmosphere, Monteiro (1976, p. 100) divided them into three abstract channels, thermal comfort, air quality and meteorological impact, in order to facilitate studies related to the urban climate. The thermal comfort channel is related to the energy balance, evidenced in the generation of the heat island, the air quality channel refers to the pollution generated in the urban area, mostly by industries and combustion vehicles, and the meteorological impact channel is made up of special types of weather, capable of producing intense precipitation.

In relation to climate scales, Monteiro (apud CASTRO, 1995, p. 10) "establishes links between them and urban areas, suggesting that the regional scale covers the megalopolis; the local scale involves the metropolis; the mesoclimatic scale encompasses the large city or the neighborhood of a metropolis; the topoclimatic scale involves the small city or just parts of a neighborhood; and the microclimatic scale covers a dwelling or a housing sector. This shows the urban climate on both the regional scale (large urban agglomerations) and the local scale (large, medium and small cities), as well as the levels within the local scale (mesoclimate, topoclimate and microclimate)".

Ayoade (1986, p. 300), in turn, points out that the city, due to the processes of urbanization and industrialization, has significant chemical alterations in its atmosphere. He thus concludes that the greatest impacts caused by man on the climate occur in urbanized areas, stating that,

"Man has exerted such a great impact on these areas that the urban climate is quite distinct in its characteristics from the climate of the surrounding rural areas." (AYOADE, 1986, p. 300)

Among the many changes observed in the urban climate, the author highlights alterations in the thermal and hydrological properties of the earth's surface, as well as its aerodynamic parameters, a reduction in longwave and shortwave radiation due to surface paving, an increase in temperatures, precipitation, cloudiness and the frequency of fog and mist.

It is undeniable that the climate of cities is extremely altered in terms of energy, heat and water balance, which has significant repercussions on the physical and chemical properties of their atmosphere. These changes occur due to various factors, such as the density of the built-up area, the constant release of gases by motor vehicles and industries, and the sealing of roads. Among the main changes are the increase in heat, air pollution and, of particular note, the increase in rainfall and the frequency of storms.

According to Ayoade (1986, p. 309), the increase in precipitation is partly due to the abundant presence of condensation nuclei from the pollutants generated by motor vehicles and industries. He also considers other factors, such as the increase in water vapor due to combustion processes, thermal convection over the urban heat island and the mechanical turbulence created by the friction effect of city structures on air flows.

In short, the urban environment, as Drew (1994, p. 181) puts it, has a real "climate vault" of its own, within which all its characteristics differ to some extent from the dominant regional climate. In addition, he states that in a large industrialized urban area, annual rainfall can change by 5% and days of minimum rainfall by 10%.

III - GEOPHYSICAL ASPECTS OF THE STUDY AREA

The municipality of Rio Claro (SP) covers an area of 498,422 km^2 , located at 22.41° latitude and 47.56° longitude and is home to 186,253 inhabitants (IBGE, 2013). It is bordered by the municipalities of Corumbatai and Leme (to the north), Araras and Santa Gertrudes (to the east), Piracicaba (to the south) and Ipeùna and Itirapina (to the west). Altitudes in most of the city range from 600 to 625 meters.

It is part of the Campinas Administrative Region and has a Human Development Index (HDI) of 0.803. Its urbanization rate is 97.60% (IBGE, 2004).

In terms of its geomorphology, according to Almeida's classification (1974, p. 20) it is part of the Paulista Peripheral Depression, with slightly convex tabular hills and gently sloping plateaus.

Geologically, most of the city is located on the interfluve formed by the Rio Claro Formation, while the rest lies on the Corumbatai Formation, and the valleys are formed by Holocene sediments.

The predominant soils are: red-yellow latosol on the interfluve, dark red latosol on the slopes of the Ribeirao Claro and the Corumbatai River and hydromorphic in the river valleys. The biomes present are the Cerrado and the Atlantic Forest.

According to Monteiro's (1973, p. 123) classification of the state of São Paulo, the city is located in the Central-Northern climate zone, and its climatic dynamics are controlled by equatorial and tropical masses, with a predominance of alternately dry and humid tropical climates. According to the Koppen classification (apud MENDONÇA, 2007, p. 120), the climate is characterized by summer rains and hot summers, known as Cwa.

Below is a map showing the location of the municipality of Rio Claro (SP) - Figure 1 - and its respective urban area:

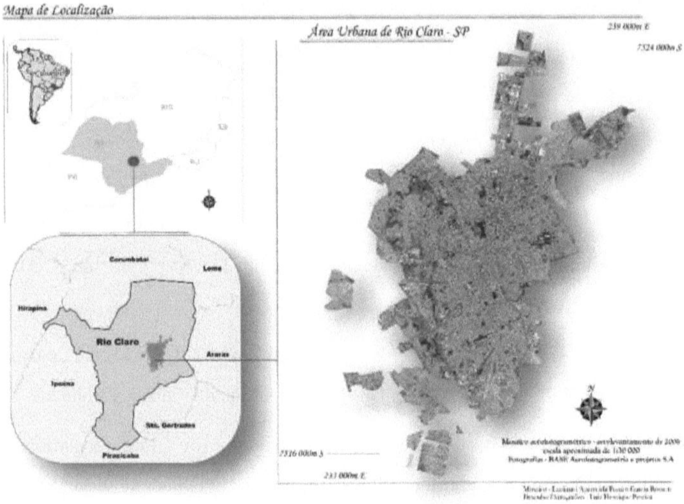

Figure 1 - Map of the Municipality of Rio Claro (SP). Source: Research Group in Territorial Analysis and Planning - GPAPT. IGCE - UNESP - Rio Claro Campus. Available at: <http://www.rc.unesp.br/igce/planejamento/gpapt/dados%20e%20imagens/mapa_Localizacao_Rio%20Claro%20_SP.jpg>

IV - OBJECTIVE

General:

- To analyze the impacts of rainfall on the population in the city of Rio Claro (SP) between 2005 and 2010.

Specifics:

- Mapping of occurrences in each area of the city of Rio Claro (SP) linked to heavy rainfall events;
- Analysis of the atmospheric systems involved in the events;
- To assist the Civil Defense authorities, the Fire Department and the executive branch in improving emergency activities and municipal planning;
- To contribute to studies related to the theme of Urban Climate, with regard to precipitation and its respective impacts on the population.

V - MATERIALS AND METHOD

This study was based on the hypothetical-deductive method, i.e. the construction of hypotheses to deduce results and predictions that can be confirmed or refuted.

As initially proposed, the project aims to analyze rainfall impacts in the urban area of Rio Claro (SP), covering the quantity, frequency and locations of occurrences, correlating them with the rainfall rates recorded and the atmospheric systems at work.

In order to carry out the mapping, news about the rains and their impact on the city was first obtained from two newspapers, Diàrio do Rio Claro and Cidade. They were collected from the Rio Claro Public and Historical Archive, resulting in a table with the dates of the events, the type of impact and the corresponding locations. However, many of the news reports did not contain the exact locations of the incidents. Therefore, it was very important to check the occurrence reports with the Fire Brigade, because the locations were more detailed.

Based on the information collected above, it was possible to map the incidents using the *CorelDRAW* program. In total, there were seven maps, one referring to the critical points in terms of the frequency of occurrences, and the other six corresponding to the locations and types of impacts for each year of the period analyzed.

Daily rainfall data from 2005 to 2010 was collected at the Center for Environmental Analysis and Planning (CEAPLA), located on the campus of UNESP, Rio Claro. In addition to the daily rainfall, the bulletins obtained contained the monthly total and the annual total. The Climatology Laboratory of the Department of Geography, part of the Institute of Geography and Exact Sciences of UNESP, Rio Claro campus, also collected daily rainfall data from weather stations located in different parts of the city.

By adding the rainfall data to the rest already acquired, it was possible to infer which period of the year the occurrences were concentrated, the months with the highest rainfall indices and the vulnerability of some locations, since even not so high rainfall caused several occurrences.

With regard to data on atmospheric systems, the synoptic charts for the days required for the research were collected from the library of the National Institute for Space Research (INPE) in Cachoeira Paulista and from the Brazilian Navy website.

Graphs were then drawn up using the *Excel* program, with the aim of relating the different types of impact to the four types of atmospheric system found: convective rainfall, rainfall associated with the entry of a frontal system, rainfall associated with the induced trough and rainfall associated with the South Atlantic Convergence Zone (SACZ). There were a total of eight graphs, which made it possible to see in which years a particular atmospheric event was most active and on which impacts it had the greatest influence.

In order to organize all the data collected, spreadsheets were created for each year of the study period in *Excel*, making it easier to analyze the information together. The spreadsheets were then transferred to the *Word* program to be attached to this paper.

VI - RESULTS AND DISCUSSIONS

Based on the data obtained from the news reports in the two newspapers analyzed and the Fire Department's incident reports, 13 types of impact were observed during the period studied: flooding, tree falls, damage to roads, damage to property, landslides, landslides, problems with the sewage system, problems with the water supply, power outages, damage to the telephone system, damage to erosion works, mud and death. Below are the numbers of incidents related to the impacts in each year analyzed:

Table 1 - Number of occurrences of each type of impact in the period 2005 to 2010

	2005	2006	2007	2008	2009	2010
Flooding	32	18	14	17	14	38
Falling tree	16	5	3	5	6	8
Road damage	8	2	7	2	-	10
Property damage	-	-	-	-	10	1
Collapse	8	1	-	1	1	1
Sliding	-	-	-	-	-	3
Problems with the sewage system	1	1	-	-	1	3
Problems with the water supply	1	-	-	-	-	1
Lack of energy	1	-	-	-	-	-
Damage to the telephone system	-	-	-	1	-	-
Damage to erosion works	-	-	-	-	-	1
Mud	-	-	-	2	-	2
Death	-	-	-	-	-	1

Source: Organized by LOPES, 2012

In total, there were 247 incidents, especially flooding, which together accounted for 133. The other two most common types of incident were falling trees (43) and damage to roads (29). The least common impacts, with only one occurrence each, were: power failure, damage to the telephone system and death.

With regard to the locations, below is a map (Figure 1) of the critical points in terms of the frequency of occurrences during the period studied:

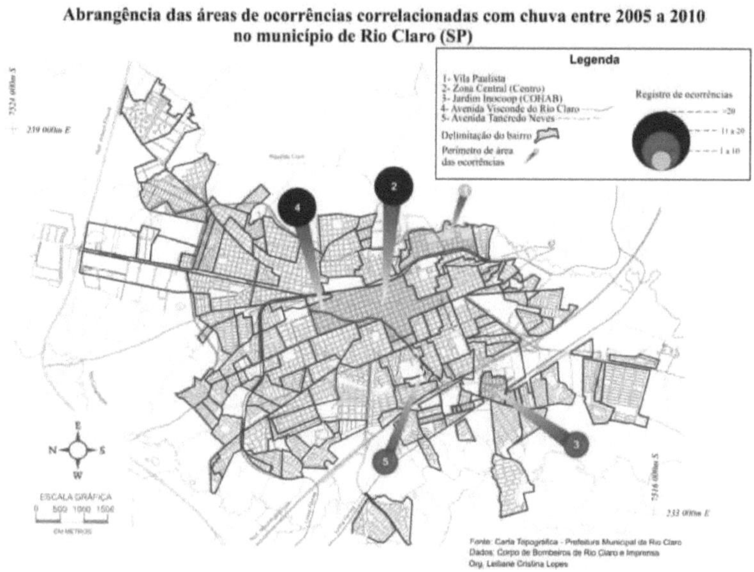

Figure 1

Among the locations highlighted above, Avenida Visconde do Rio Claro and Zona Central have the highest occurrence rates, with a total of 23 each. The Jardim Inocoop neighborhood, on the other hand, has 20, followed by Avenida Tancredo Neves, with 12 and Vila Paulista, with 10. It is thus possible to notice a concentration of occurrences in the central area of the city, with the exception of the Jardim Inocoop and Vila Paulista neighborhoods.

However, although the Jardim Inocoop neighborhood is located more peripherally, it is the continuation of the course of the Servidao stream, which overflows and is one of the factors responsible for the flooding that occurs on Visconde do Rio Claro and Tancredo Neves avenues, as well as in Jardim Inocoop.

In relation to the daily rainfall figures, below are the days that recorded rainfall above 50 mm and had the highest number of occurrences:

Table 2 - Amount of rainfall and number of occurrences

Day	Precipitation (mm)	Number of occurrences
22/01/2005	64,7	11
29/01/2005	122,6	14
23/11/2005	78,4	17
20/11/2006	56,5	10
18/01/2007	54,9	9
08/01/2010	106,9	25
25/02/2010	61	15

Source: Organized by LOPES, 2012

An analysis of the table above shows that the days with the highest number of occurrences, as well as the highest rainfall rates, belong to the months of January, February and November. In this way, it became clear that the occurrences were concentrated in the spring-summer period during the six years studied, due to the fact that it is characterized as the rainy season of the climate corresponding to the location of the municipality of Rio Claro. According to Monteiro's (1973, p. 123) classification of the state of São Paulo, the city is located in the Central-Northern climate zone, i.e. on the São Paulo Peripheral Depression, and its climatic dynamics are controlled by equatorial and tropical masses, with a predominance of alternately dry and humid tropical climates.

Therefore, of the 247 incidents, 233 occurred in the spring-summer period, with January standing out as the month that recorded the highest rainfall rates in 2005, 2007 and 2010 and, consequently, the highest number of incidents in a single day.

Analysis of the synoptic charts corresponding to the days on which the events were recorded showed a predominance of convective rainfall, 49% of the total, as can be seen in the following graph (Graph 1):

Gráfico 1 - Quantidade de ocorrências relacionadas aos sistemas atmosféricos entre os anos de 2005 e 2010

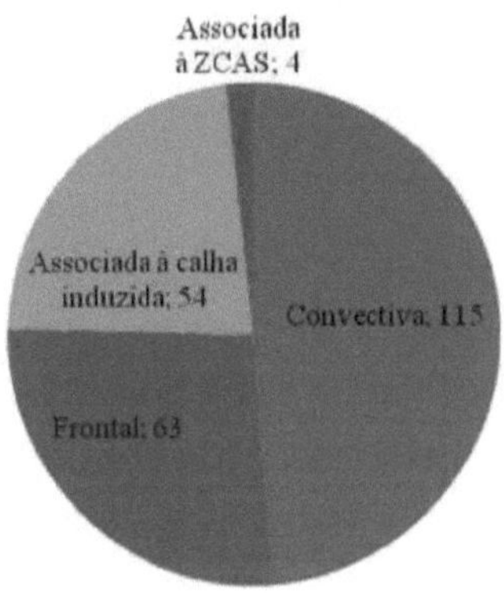

Source: Organized by LOPES, 2012

Rainfall associated with the entry of a frontal system appears as the second most frequent type of atmospheric system (26%), followed, with little difference, by rainfall related to the induced trough (23%). Rainfall related to the South Atlantic Convergence Zone (SACZ) is the least frequent, accounting for only 2% of the total. The next graph (Graph 2) elucidates the relationship between the types of occurrences and which atmospheric systems were associated with them during the period analyzed.

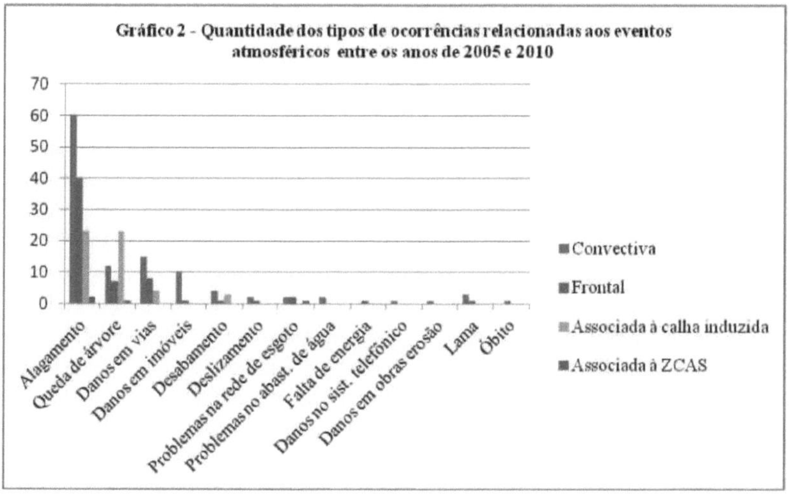

Source: Organized by LOPES, 2012

It was observed that convective rainfall was the most recurrent in the vast majority of occurrences. However, with regard to the impact corresponding to falling trees, there is a predominance of rain associated with induced rainfall. The impact of collapses is equally related to convective rainfall and induced rainfall. The occurrences of power failure and death were caused by rainfall linked to the entry of a frontal system.

In order to visualize and analyse the location of the different types of impact, the occurrences for each year of the period studied were mapped using the tables (attached) containing the data acquired for the research. Below is the map for 2005 (Figure 2):

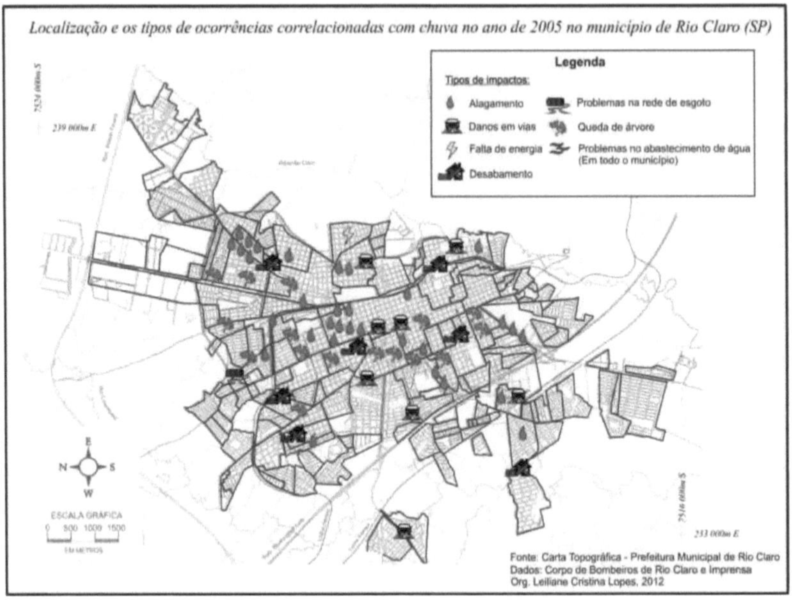

Figure 2

The year 2005 (table 3) saw a high number of incidents, totaling 67. Among them were: 32 floods, 16 tree falls, 8 road damages, 8 landslides, one power failure, one related to problems with the sewage system and one to problems with the water supply. The flooding was mostly concentrated in Visconde do Rio Claro, Ulisses Guimaraes and Presidente Kennedy avenues. With regard to falling trees, although there were incidents in more peripheral neighborhoods, the central area of the city was the hardest hit. It can be said that the other types of impact were distributed in different locations, without any concentration in any specific area.

With regard to rainfall, the months of January and November recorded high rainfall indices and together accounted for more than 50% of the occurrences in the year. It was also possible to observe that rainfall of around 15 and 20 mm has already caused occurrences, as in the cases of September 16, which recorded only 20.5 mm and December 16, with 24.2 mm.

The atmospheric systems were analyzed in order to correlate the type of event with the different impacts observed. Graphs were therefore drawn up for each year studied in order to facilitate correlations. Below is the graph (Graph 3) for 2005:

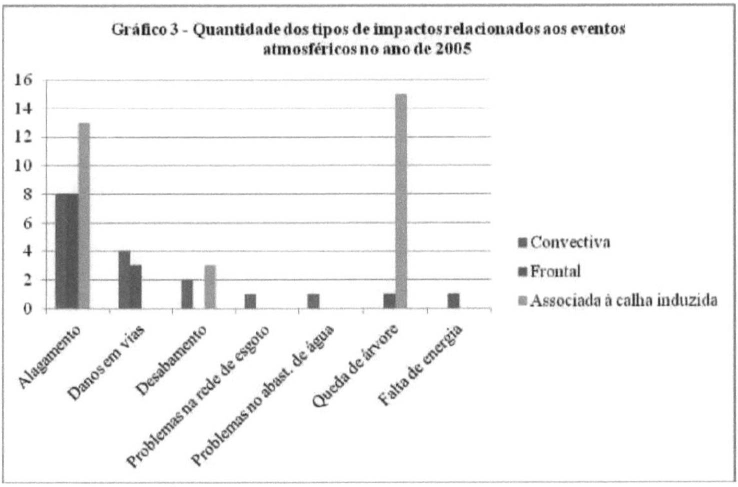

Source: Organized by LOPES, 2012

First of all, before starting the analysis with the graphs, it is necessary to point out that the totals of occurrences in graphs 3 and 4 do not correspond to the actual amount verified, due to the fact that two synoptic charts, from September 16, 2005 and October 12, 2006, could not be found, either on the Navy website or in the INPE library in Cachoeira Paulista.

From the graph above, it can be seen that in 2005 there was a marked presence of rainfall associated with the induced trough, amounting to a total of 31 occurrences, while those related to convective rainfall amounted to only 16. With regard to frontal rainfall, this was responsible for 13 occurrences, mostly flooding. No occurrences associated with the SACZ were observed this year.

In 2006 (Table 4) there was a significant decrease in the number of occurrences compared to the previous year, as can be seen in the map below (Figure 3):

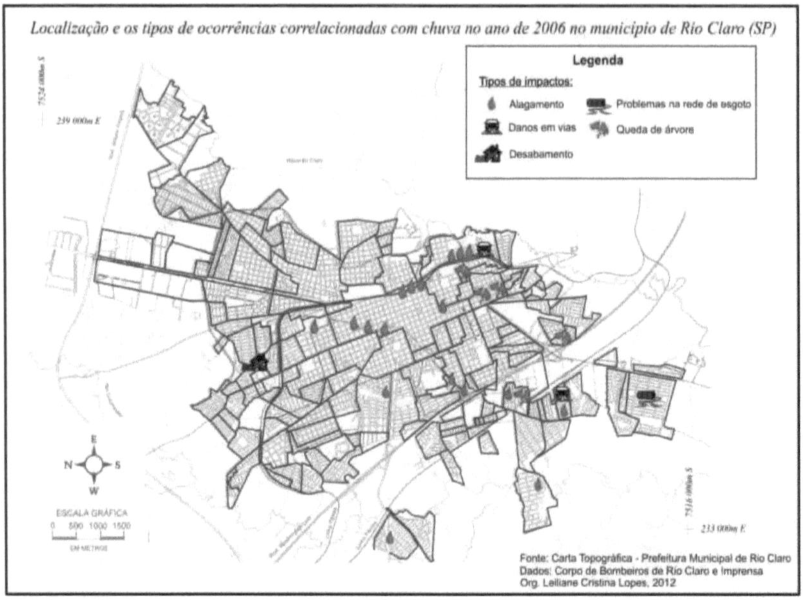

Figure 3

In total, there were 27 incidents, most of which were flooding, which amounted to 18. Among the other types of impact were: 5 fallen trees, 2 damages to roads, one collapse and one related to problems with the sewage system. This year, it was noted that in addition to Visconde do Rio Claro and Tancredo Neves avenues, there was a concentration of flooding at the Avenida 29 bridge, in the Vila Paulista neighborhood, and at the Avenida 7 and Rua 1 bridge, in the Central Zone. It was also noted that three of the trees fell near the Aero Clube.

This year's rainfall was lower than in 2005, except for the month of November, which recorded 56.5 mm, leading to 10 floods in different parts of the city.

In relation to the atmospheric systems at work, the following graph is shown (Graph 4):

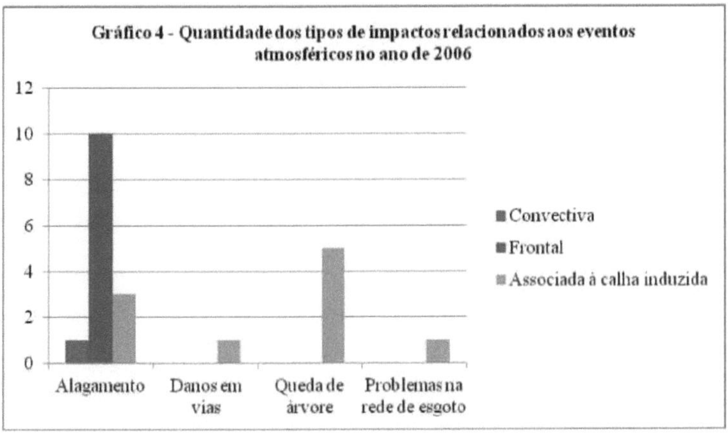

Source: Organized by LOPES, 2012

Although most of the flooding was caused by frontal systems, the occurrences in 2006 were equally influenced by frontal rainfall and rainfall associated with the induced trough, both with a total of 10 occurrences. Only one of the floods was caused by convective rain and, like the previous year, there were no impacts associated with the SACZ.

The year 2007 (Table 5) had the lowest number of occurrences, as shown in the map below (Figure 4):

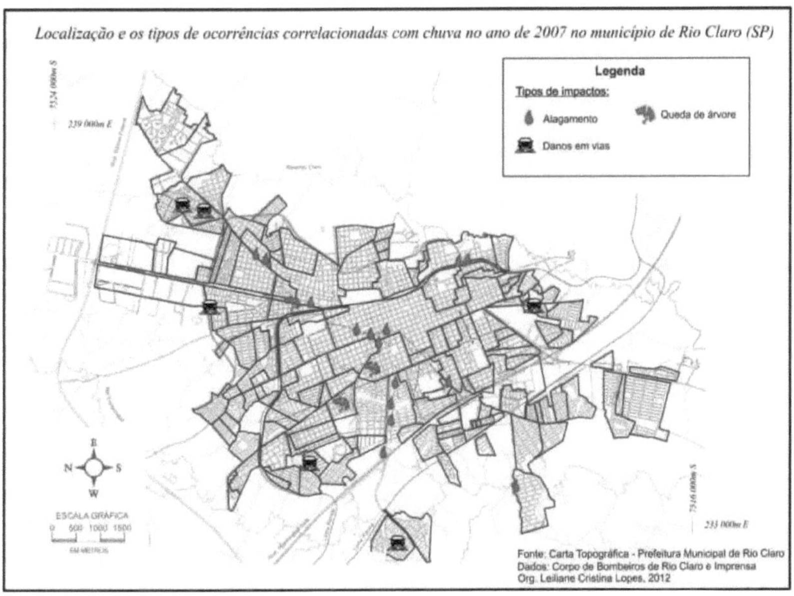

Figure 4

A total of 24 incidents were recorded, of which 14 were flooding, 7 were road damage and 3 were fallen trees. In addition to the flooding points mentioned above, this year there was a concentration on Avenida Castelo Branco. It was also possible to observe that the incidents of damage to roads predominated in more peripheral areas of the city.

Regarding rainfall, although January was the month with the most occurrences, the highest rainfall was recorded on October 24, with 82.2 mm.

Below is a graph showing the predominant atmospheric events (Graph 5):

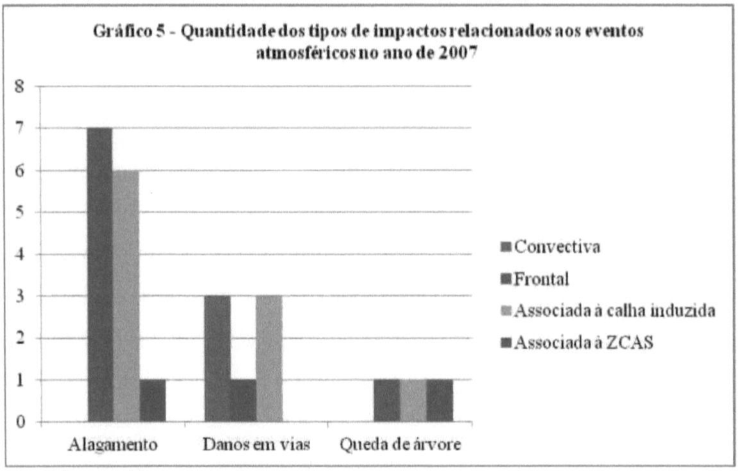

Source: Organized by LOPES, 2012

Again, according to the analysis of the graph, there were few incidents of convective precipitation, only 3 of damage to roads. This year, the atmospheric systems that predominated were those associated with the induced trough, with 10 occurrences, and the entry of fronts, with a total of 9. There were also two occurrences associated with the SACZ.

Below is the map (Figure 5) for 2008 (Table 6):

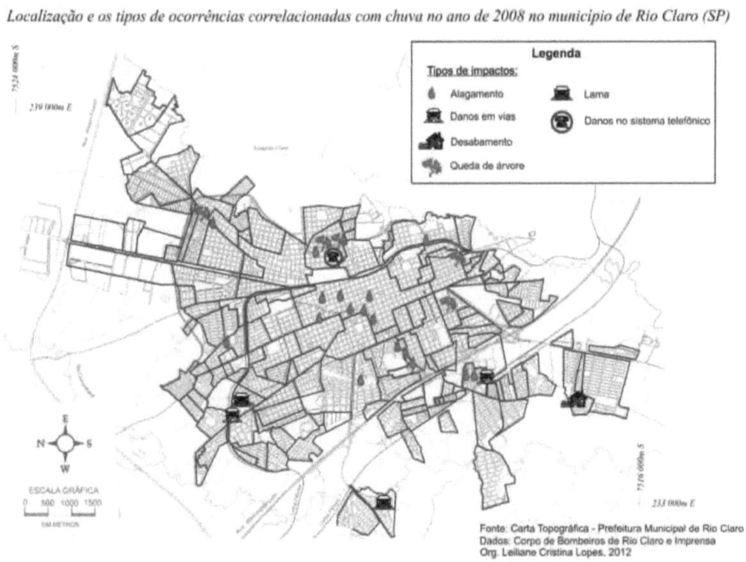

Figure 5

This year, there were 28 incidents, of which 18 were flooding, 5 were falling trees, 2 were damage to roads, 2 were mud and one was damage to the telephone system. Looking at the map, you can clearly see the concentration of flooding along the course of the Córrego da Servidao, which corresponds to Visconde do Rio Claro and Tancredo Neves avenues and the Jardim Inocoop neighborhood. As already mentioned, the damage to roads was located in the more peripheral neighborhoods, as was the occurrence of mud, due to the fact that some of these neighborhoods are unpaved. Although some tree falls were random, two of the incidents happened near the Aero Clube.

In terms of rainfall, precipitation of around 20 mm has already caused several incidents, especially at the points considered most problematic in terms of the frequency of impacts. The months with the highest number of incidents were February, October and December.

With regard to atmospheric systems, only two types were observed in 2008, as shown in the following graph (Graph 6):

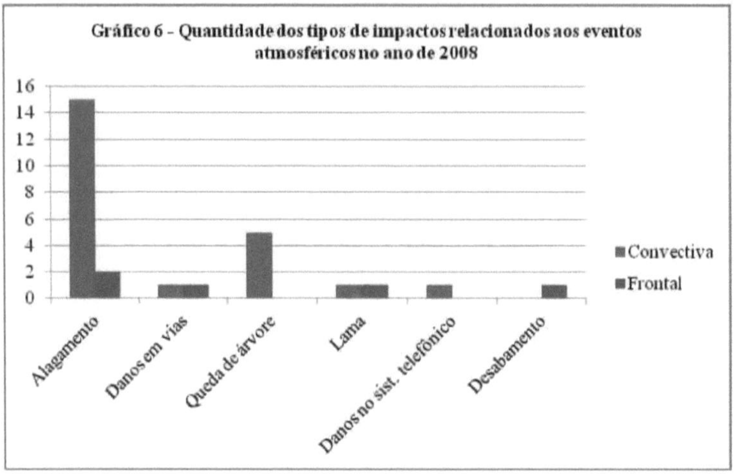

Source: Organized by LOPES, 2012

It was clear that the predominant atmospheric system was that related to convective precipitation, with a total of 23 occurrences, leaving only 5 related to the entry of frontal systems. There were no occurrences corresponding to the induced trough or the SACZ.

The year 2009 (Table 7) saw a not very significant increase in the number of occurrences, as can be seen in the map below (Figure 6):

Figure 6

In total, there were 32 incidents, including: 14 floods, 10 damages to property, 6 fallen trees, one collapse and one related to problems with the sewage system. This year, there was an increase in the number of floods in the Jardim Inocoop neighborhood. There was also a concentration of this type of impact on Avenida Castelo Branco. Tree falls, on the other hand, were more random than in previous years.

The month of December was significantly rainy, with 119.7 mm recorded on the 29th, making it the month with the highest number of occurrences. The month of February, although it didn't record such high rainfall indices, also had a large number of occurrences.

Below is a graph of the predominant atmospheric systems (Graph 7):

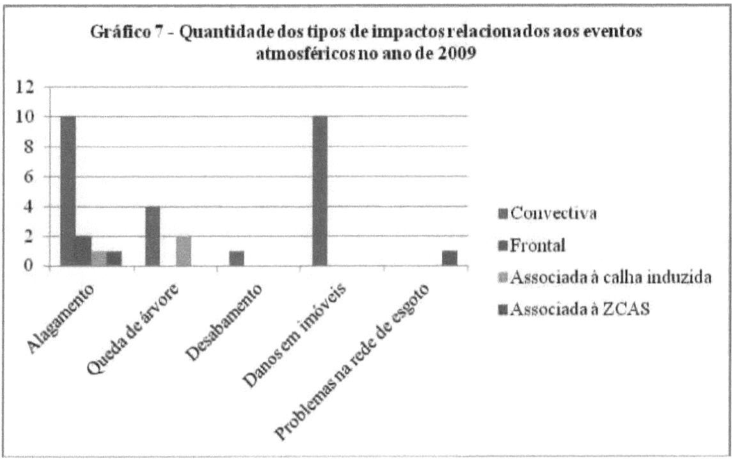

Source: Organized by LOPES, 2012

It can be inferred from the graph above that most of the occurrences are related to convective rainfall, as was the case the previous year, with a total of 25. The other types of atmospheric system played a lesser role, with a total of only 7 occurrences.

The last year of the study period, 2010 (Table 8), had the highest number of occurrences, as shown in the map below (Figure 7):

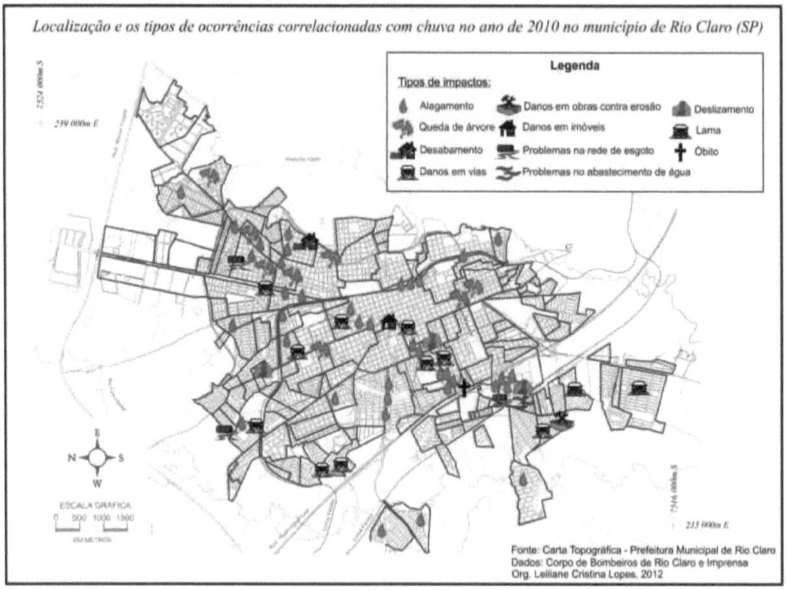

Figure 7

In addition to the large number of incidents, totaling 69, there was also a greater variety of types of impact, including: 38 floods, 10 damages to roads, 8 tree falls, 3 landslides, 3 related to problems with the sewage system, 2 related to mud, one landslide, one corresponding to damage to property, one to damage to erosion works, one to problems with the water supply and one to death. Flooding tended to be concentrated at the points already analyzed, with only a few exceptions. Damage to roads, although still mostly observed in outlying neighborhoods, there were four occurrences in the central area. Unlike in previous years, tree falls were more prevalent in the north of the city, with only two occurrences near the Aero Clube. This year, there was one death on Avenida Tancredo Neves, near the access to the Washington Luis highway.

Of the 69 incidents during the year, 68 occurred in January, February and December, making for an extremely rainy summer. One of the heaviest rainfalls on record, 173.8 mm, occurred on February 25, causing numerous floods in various neighborhoods and avenues, as well as the death.

With regard to atmospheric systems, see the graph below (Graph 8):

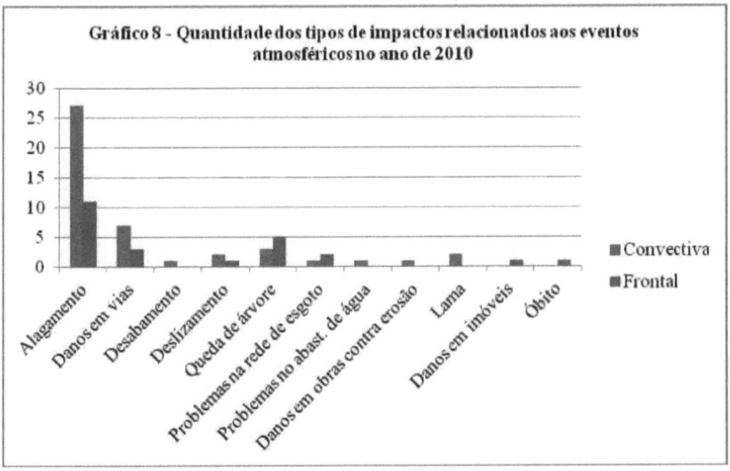

Source: Organized by LOPES, 2012

Although 2010 had the highest number of occurrences, 45 of the total were associated with convective rainfall, while only 24 were caused by frontal systems. However, the highest rainfall index of the year was caused by frontal rain. There were no occurrences corresponding to the induced trough or the SACZ.

VII - CONCLUSIONS

The mapping showed that the critical points in terms of frequency of occurrences are: Avenida Visconde do Rio Claro, Avenida Tancredo Neves, the Central Zone and the Jardim Inocoop and Vila Paulista neighborhoods. Although there was randomness in most of the types of impact, flooding, falling trees and damage to roads showed a pattern in terms of location.

Among the locations where flooding is most frequent are Avenida Visconde do Rio Claro and Avenida Tancredo Neves, both of which experienced flooding in all the years analyzed. This type of impact was also very present in the Jardim Inocoop neighborhood, with a considerable increase in the number of occurrences in 2009 and 2010. In the Castelo Branco and José Felicio Castellano avenues and on the bridges of Avenida 7, Zona Central and Avenida 29, Vila Paulista, flooding is also very common. It was also noted that at these points where it is more common for impacts to occur, rainfall of 20 mm or more usually causes flooding, highlighting the vulnerability of these locations.

With regard to the occurrences of falling trees, there were almost no areas that could be considered recurring sites. However, there was a concentration of this type of impact in places close to the Aero Club, which may be associated with the fact that it is an open field, facilitating the passage of strong winds.

With regard to the impacts corresponding to damage to roads, the occurrences prevail in the city's more peripheral neighborhoods. Occurrences related to mud are also located in outlying areas, due to the lack of paving.

It is not possible to make spatial inferences regarding the occurrences of the rest of the verified impacts, as they are found randomly, without any kind of pattern.

Based on the analysis of the graphs, although the most recurrent atmospheric system during the period studied was convective precipitation, the analysis of each year separately showed that the predominance of the different types of atmospheric events varied. However, there is the exception of the system associated with the South Atlantic Convergence Zone (SACZ), which is related to only four occurrences, thus not predominating in any of the years analyzed.

REFERENCES

ALMEIDA, F. F. M. de. *Geological foundations of the São Paulo relief.* Sao Paulo: IGEOG- USP series of theses and monographs No. 14, 1974. 99 p.

AYOADE, J. O. *Introduction to climatology for the tropics.* Sao Paulo: DIFEL, 1986. 332 p.

CASTRO, A. W. S. *Urban Climate: Rainfall in Rio Claro - SP.* 1995. 196f. Dissertation (Master's Degree in Space Organization) - Institute of Geosciences and Exact Sciences, Universidade Estadual Paulista "Júlio de Mesquita Filho", Rio Claro. 1995.

DREW, D. *Processos interativos homem-me ambiente.* Rio de Janeiro: Bertrand Brasil, 1994. 224 p.

Brazilian Institute of Geography and Statistics (IBGE) - "Cidades@". Available at: <http://www.ibge.gov.br/cidadesat/default2.php>. Accessed on Aug. 15, 2012.

MENDONÇA, F.; DANNI-OLIVEIRA, I. M. *Climatologia - Noções basics do clima no Brasil.* Sao Paulo: Oficina de Textos, 2007. 208 p.

MONTEIRO, C. A. F. A *dinàmica climàtica e as chuvas no estado de São Paulo, estudo geogràfico sob a forma de atlas.* Sao Paulo: USP - Institute of Geography, 1973. 130 p.

C. A. F. *O clima e a organizaçãoQão do espaço no estado de São Paulo: problemas e perspectivas.* Sao Paulo: IGEOG-USP series of theses and monographs No. 28, 1976. 54 p.

C. A. F. *Teoria e clima urbano.* Sao Paulo: IGEOG-USP series of theses and monographs No. 25, 1976. 181 p.

MONTEIRO, C. A. F.; MENDONÇA, F. *Clima Urbano.* Sao Paulo: Contexto, 2011. 192 p.

SANTOS, M. *A urbanização brasileira.* Sao Paulo: Hucitec, 1993. 176 p.

M. *Manual de geografia urbana.* Sao Paulo: Hucitec, 1981. 232 p.

SPOSITO, M. E. B. *Capitalismo e Urbanizaçâo.* Sao Paulo: Contexto, 1996. 80 p.

ZAVATTINI, J. A. *Estudos do clima no Brasil.* Campinas: Alinea, 2004. 398 p.

ANNEXES

ANNEX A - Record of the amount of rainfall and its relationship with the location and types of occurrences and the atmospheric systems active in 2005

January						
Date	**Type of Occurrence**	**Location**	**Precipitation (mm)**	**Rainfall station near the town (mm)**	**Obs:**	**Type of Atmospheric System**
22	Flooding	Avenida Visconde do Rio Claro x 8th Street - Center	64,7	11.7 (Fepasa)	02h00m to 03h00m	Precipitation associated with the entry of a frontal system
	Flooding (home)	Avenue M- 17A - Jardim Cherveson (n° 1298)		38.6 (Cavalry)		
	Flooding	Ulisses Guimaraes Avenue		11.7 (Fepasa)		
	Flooding	Vila Cristina		54 (Brastemp)		
	Flooding	Vila Paulista		44.6 (Gessy Ineos)		
	Flooding	Vila Operària		9 (Santana)		
	Flooding	Avenida Presidente Kennedy		44.6 (Gessy Ineos)		
	Road damage	Ulisses Guimaraes Avenue		11.7 (Fepasa)		
	Road damage	Avenue Visconde do Rio Claro		11.7 (Fepasa)		
	Road damage	Av. 29 bridge - Vila Paulista		44.6 (Gessy Ineos)		
	Lack of energy	Vila Bela Vista		57.2 (DAAE Horto)		
29	Flooding	Avenue Visconde do Rio Claro	122,6	81.4 (Fepasa)	03h00 to 09h00m	Convective rainfall
	Flooding	Felicio Castellano Avenue		150 (Brastemp)		
	Flooding	Avenida Presidente Kennedy		117.6 (Gessy Ineos)		
	Flooding	Jardim Inocoop		135 (Nheel Química)		
	Flooding	Tancredo Neves Avenue		135 (Nheel Química)		
	Flooding	Avenida Brasil		132.8 (Tigre)		
	Flooding	Garden Wenzel		86.4 (Cavalry)		
	Road damage	Jardim Inocoop		135 (Nheel Química)		
	Road damage	Av. 4 x Rua 7 e 8 - Centro		81.4 (Fepasa)		
	Damage to roads (cycle paths)	Between Jardim Bom Sucesso and Jardim Novo Wenzel		135 (Nheel Química)		
	Collapse	Av. 11 wooden		135 (Nheel		

		bridge		Química)		
	Collapse	Access bridge to Jardim Nova Rio Claro		135 (Nheel Química)		
	Problems with the sewage system	Cherveson Garden		86.4 (Cavalry)		
	Problems with the water supply	The whole municipality		-		
March						
Date	**Type of Occurrence**	**Location**	**Precipitation (mm)**	**Rainfall station near the location (mm)**	**Obs:**	**Type of Atmospheric System**
14	Flooding	Avenue Visconde do Rio Claro	43,2	32.9 (Fepasa)	22h00m to 24h00m	Precipitation associated with the entry of a frontal system
May						
Date	**Type of Occurrence**	**Location**	**Precipitation (mm)**	**Rainfall station near the location (mm)**	**Obs:**	**Type of Atmospheric System**
24	Flooding	Avenue Visconde do Rio Claro x	72	79.8 (Fepasa)	23h00m to 24h00m	Convective rainfall
		11th Street				
	Road damage	Crossroads of Avenues 8 and 10		79.8 (Fepasa)		
September						
Date	**Type of Occurrence**	**Location**	**Precipitation (mm)**	**Rainfall station near the town (mm)**	**Obs:**	**Type of Atmospheric System**
16	Flooding	Downtown streets and avenues	20,5	53.2 (DAAE Horto)	12h00 to 13h00m	-
	Flooding (commerce)	Bus station restaurant		31.6 (Nheel Quimica)		
	Flooding (home)	Avenida 7 - Jardim Res. das Paineiras		13.4 (Agroceres)		
	Road damage	Rua 14 x Av. 22 and 24 - Santa Cruz		29 (Santana)		
	Collapse	Jardim Olinda		9.1 (ETE Flores)		
	Collapse	Rua 14 x Av. da Saudade - Bairro da Consolaçâo		40 (Gessy Ineos)		
	Collapse	Av. 60 - University Park		9.1 (ETE Flores)		
November						
Date	**Type of Occurrence**	**Location**	**Precipitation (mm)**	**Rainfall station near the location (mm)**	**Obs:**	**Atmospheric System Type**
	Flooding	Avenue Visconde do Rio Claro		33.2 (Fepasa)		
	Flooding	Rua 1 x Av. 26 - Center		33.2 (Fepasa)		
	Flooding	Unidade Bàsica de Saùde - Vila Cristina		12 (Chemson)		
	Flooding	Av. Presidente		55.5 (Gessy Ineos)		

		Kennedy - Rotatòria in				
23		in front of the 3 farms	78,4		3:30 p.m. to 16h10m	Precipitation associated with the induced trough
	Flooding	Rua 5 x Av. 11 - Centro		67.9 (DAAE Horto)		
	Flooding	Pontilhao da Av. 7 x Rua 1 - Centro		67.9 (DAAE Horto)		
	Flooding	Pontilhao da Av. 15 x Av. Paulista - Jardim Primavera		86 (Santana)		
	Flooding	Felicio Castellano Avenue		12 (Chemson)		
	Flooding	Rua 11 A - Vila Nova		33.2 (Fepasa)		
	Flooding	Ulisses Guimaraes Avenue		33.2 (Fepasa)		
	Flooding	Rua 5JA x Av. 54A - Vila Cristina		12 (Chemson)		
	Flooding / Collapse	Efraim School Ranch Ribeiro - Av. José Felicio Castellano - Vila Cristina		12 (Chensom)		
	Collapse	Francisco Matarazzo Avenue - Vila Paulista		55.5 (Gessy Ineos)		
	Collapse (Lining - Shopping Center)	Av. 24A x Rua 6 - Santana		86 (Santana)		
	Falling tree	Av. 38 x Rua 5A and 4A - Vila Alema		33.2 (Fepasa)		
	Falling tree	Street M-04 - Jardim Hipódromo		42.9 (ETE Flores)		
24	Falling tree	Street 3A - Arco-iris	11	1.4 (Chemson)	21h00m	Precipitation associated with the induced trough
	Falling tree	16th Street - Jardim Claret		13.4 (Gessy Ineos)		
	Falling tree	Av. 58A - Jardim América		1.4 (Chemson)		
	Falling tree	Av. 68 - Jardim		14.4 (Cavalry)		
		Olinda				
	Falling tree	Av. 1 - Center		10 (Fepasa)		
	Falling tree	Cherveson Garden		14.4 (Cavalry)		
25	Falling tree	Street 1A - Vila do Ràdio	15,2	13.4 (Gessy Ineos)	11:00 a.m. to 12h00 and at 15h00m	Precipitation associated with the entry of a frontal system
December						
Date	**Type of Occurrence**	**Location**	**Precipitation (mm)**	**Rainfall station near the location (mm)**	**Obs:**	**Type of Atmospheric System**
16	Flooding	Avenue Visconde do Rio Claro	24,2	20.5 (Fepasa)	14h00 to 16h00m	Precipitation associated with the induced trough
	Falling trees (2)	Av. 23 - Estàdio district		23.3 (Gessy Ineos)		

	Falling trees (2)	Rua 13 x Av. 12 and 14 - Santa Cruz		29 (Santana)		
	Falling tree	Rua 3 x Av. 42 and 46 - Vila Operària		29 (Santana)		
	Falling tree	Santana		29 (Santana)		
	Falling tree	Rua 3 x Av. 42 and 46 - Vila Operària		29 (Santana)		

Source: Organized by LOPES, 2012.

ANNEX B - Record of the amount of rainfall and its relationship with the location and types of occurrences and the atmospheric systems operating in 2006

January

Date	Type of Occurrence	Location	Precipitation (mm)	Rainfall station near the location (mm)	Obs:	Atmospheric System Type
10	Flooding	Visconde do Rio Claro Avenue	3,9	0 (Fepasa)	13h00m to 14h00m	Convective rainfall

October

Date	Type of Occurrence	Location	Precipitation (mm)	Rainfall station near the location (mm)	Obs:	Atmospheric System Type
12	Flooding	Blue Lake - Vila Operària	35,1	57 (Santana)	05h30m to 06h30 and at 15h00m	-
	Flooding	Visconde do Rio Claro Avenue		32 (Fepasa)		
	Flooding	Av. 29 bridge - Vila Paulista		73.6 (Gessy Ineos)		
	Flooding	Pontilhao da Av. 7 x Rua 1 - Centro		34 (DAAE Horto)		
	Road damage	Rua 2 x Av. 4 - Jardim Guanabara II		36.7 (Agroceres)		
	Collapse	Cherveson Garden		34 (Cavalry)		

November

Date	Type of Occurrence	Location	Precipitation (mm)	Rainfall station near the location (mm)	Obs:	Type of Atmospheric System
20	Flooding	Pontilhao da Tancredo Neves x Washington Luis	56,5	75.8 (Gessy Ineos)	16h00m to 16:30	Precipitation associated with the entry of a frontal system
	Flooding	Visconde do Rio Claro Avenue		67 (Fepasa)		
	Flooding	Rua 20 x Av. 16 e 18 - Parque São Paulo		15 (Santana)		
	Flooding	Rua 4 x Av. Tancredo Neves - Jardim Inocoop		75.8 (Gessy Ineos)		
	Flooding	Garden New Rio Claro		75.8 (Gessy Ineos)		
	Flooding	Jardim Novo Wenzel		15 (Santana)		
	Flooding	Av. 7 x Rua 1 - Center		69.2 (DAAE Horto)		
	Flooding	Av. Tancredo Neves near Rodoviària - Jardim Claret		75.8 (Gessy Ineos)		
	Flooding	Av. 29 bridge - Vila Paulista		75.8 (Gessy Ineos)		
	Flooding (home)	Rua 6 x Av. 13 e		69.2 (DAAE		

		15 - n° 462 - Centro		Horto)		
Decem	**bro**					
Date	**Type of Occurrence**	**Location**	**Precipitation (mm)**	**Rainfall station near the location (mm)**	**Obs:**	**Type of Atmospheric System**
19	Flooding	Pontilhao da Av. 7 x Rua 1 - Centro	12	23.3 (DAAE Horto)	7pm to 20h00m	Precipitation associated with the induced trough
	Flooding	Guanabara Garden II		29.9 (Agroceres)		
	Flooding / Damage to	29th Avenue Bridge		32.9 (Gessy Ineos)		
	routes	Vila Paulista				
	Falling tree	9th Street - Cidade Jardim		32.9 (Gessy Ineos)		
	Falling tree	Rua 5 x Av. 23 e 25 - Jardim Donângela		32.9 (Gessy Ineos)		
	Falling tree	Jardim Inocoop		29.9 (Agroceres)		
	Falling tree	Jardim Anhanguera		29.9 (Agroceres)		
	Falling tree	Garden City		32.9 (Gessy Ineos)		
	Problems with the sewage system	Garden New I		29.9 (Agroceres)		

Source: Organized by LOPES, 2012.

ANNEX C - Record of the amount of rainfall and its relationship with the location and types of occurrences and the atmospheric systems active in 2007

January						
Date	**Type of Occurrence**	**Location**	**Precipitation (mm)**	**Rainfall station near the location (mm)**	**Obs:**	**Atmospheric System Type**
3	Road damage	Avenue M-37 x Avenue M-41 - Jardim Santa Clara	17,9	13 (Brastemp)	03h00 to 11h00m	Convective rainfall
5	Road damage	Av. 1 x Rua 11 - Mae Preta	64	61.5 (Florence)	02h00m to 03h00m	Precipitation associated with the entry of a frontal system
6	Falling tree	Jardim Bela Vista	21,6	40 (Santana)	03h00 to 04h00m and 06h00 to 15h00m	Precipitation associated with the entry of a frontal system
8	Road damage	11th Street - Mae Preta	18,3	2.5 (Florence)	03h00m	Convective rainfall
	Flooding	Visconde do Rio Claro Avenue		75 (Santana)		
	Flooding	Tancredo Neves Avenue		51.9 (Claretians)		
	Flooding	Avenida José Felicio Castellano		18 (Brastemp)		
	Flooding	Rua 21 x Av. 1 - Jardim Nova Rio Claro		51.9 (Claretians)		
	Flooding	Av. Castelo Branco (exit and entrance to Washington Luis) - Jardim Sao Paulo		75 (Santana)		
18	Flooding	29th Avenue Bridge	54,9	51.2 (Gessy Ineos)	03h00m and 15h00m	Precipitation associated with the induced trough
		Vila Paulista				
	Road damage	27th Street x 54th and 56th Avenues - Jardim Paulista II		51.2 (ETE Flores)		
	Road damage	7th Street - New Jd. Wenzel		51.9 (Claretians)		
	Road damage	Rua Batista Russo - Condominio Villagio Ipanema - Jardim Botànico		45 (Civil Defense)		
February						
Date	**Type of Occurrence**	**Location**	**Precipitation (mm)**	**Rainfall station near the location (mm)**	**Obs:**	**Type of Atmospheric System**
25	Falling tree	Av. 22 x Rua 11 e	27,6	87 (Santana)	6 p.m. to	Precipitation

<table>
<tr><td></td><td></td><td>12 -n° 1280 - Santa Cruz</td><td></td><td></td><td>19h00m</td><td>associated with the induced trough</td></tr>
<tr><td>28</td><td>Flooding</td><td>Vila Cristina</td><td>5,2</td><td>6 (Brastemp)</td><td>16h00m to 17h00m</td><td>Precipitation associated with the SACZ</td></tr>
<tr><td colspan="7">March</td></tr>
<tr><td>Date</td><td>Type of Occurrence</td><td>Location</td><td>Precipitation (mm)</td><td>Rainfall station near the location (mm)</td><td>Obs:</td><td>Type of Atmospheric System</td></tr>
<tr><td>19</td><td>Falling tree</td><td>Av. Brasil - Vila Martins</td><td>37,2</td><td>16 (Santana)</td><td>24h00 to 07h00m</td><td>Precipitation associated with the SACZ</td></tr>
<tr><td colspan="7">October</td></tr>
<tr><td>Date</td><td>Type of Occurrence</td><td>Location</td><td>Precipitation (mm)</td><td>Rainfall station near the location (mm)</td><td>Obs:</td><td>Type of Atmospheric System</td></tr>
<tr><td>24</td><td>Flooding</td><td>Rua 3A - n° 1252 - Vila Alemâ</td><td>82,2</td><td>53.2 (DAAE Horto)</td><td>06h00 to 24h00m</td><td>Precipitation associated with the entry of a frontal system</td></tr>
<tr><td colspan="7">November</td></tr>
<tr><td>Date</td><td>Type of Occurrence</td><td>Location</td><td>Precipitation (mm)</td><td>Rainfall station near the location (mm)</td><td>Obs:</td><td>Type of Atmospheric System</td></tr>
<tr><td rowspan="4">4</td><td>Flooding</td><td>Rua 16 - Jardim Sâo Paulo</td><td rowspan="4">48,1</td><td>9.5 (Santana)</td><td rowspan="4">01h00 to 13h00m</td><td rowspan="4">Precipitation associated with the entry of a frontal system</td></tr>
<tr><td>Flooding</td><td>Avenue Visconde de Rio Claro</td><td>9.5 (Santana)</td></tr>
<tr><td>Flooding</td><td>Pontilhao Av. 29 - Vila Paulista</td><td>11.7 (Gessy Ineos)</td></tr>
<tr><td>Flooding</td><td>Rua 8 x Avenida 16 - Santa Cruz</td><td>9.5 (Santana)</td></tr>
<tr><td colspan="7">December</td></tr>
<tr><td>Date</td><td>Type of Occurrence</td><td>Location</td><td>Precipitation (mm)</td><td>Rainfall station near the location (mm)</td><td>Obs:</td><td>Atmospheric System Type</td></tr>
<tr><td rowspan="2">12</td><td>Flooding</td><td>Jardim Sao Paulo</td><td rowspan="2">37,1</td><td>53.2 (Claretians)</td><td rowspan="2">24h00 to 12h00m</td><td rowspan="2">Precipitation associated with the entry of a frontal system</td></tr>
<tr><td>Flooding</td><td>Castelo Branco Avenue</td><td>53.2 (Claretians)</td></tr>
<tr><td>19</td><td>Road damage</td><td>Junction of Av. 14 and Avenida Visconde do Rio Claro - Santana</td><td>42,2</td><td>51 (Santana)</td><td>13h00m to 14h00m</td><td>Convective rainfall</td></tr>
</table>

Source: Organized by LOPES, 2012.

ANNEX D - Record of the amount of rainfall and its relationship with the location and types of occurrences and the atmospheric systems active in 2008

January						
Date	**Type of Occurrence**	**Location**	**Precipitation (mm)**	**Rainfall station near the location (mm)**	**Obs:**	**Type of Atmospheric System**
20	Falling tree	Rua 4 x Av. 39 and 41 - City Garden	27,5	22 (Civil Defense)	11:00 a.m. to 16h00m	Convective rainfall
February						
Date	**Type of Occurrence**	**Location**	**Precipitation (mm)**	**Rainfall station near the location (mm)**	**Obs:**	**Atmospheric System Type**
18	Flooding	Av. Tancredo Neves - in the bus station area	53,2	52.2 (Civil Defense)	7pm to 20h00m	Convective rainfall
	Flooding	Jardim Inocoop		52.2 (Civil Defense)		
	Flooding	Ulisses Guimaraes Avenue		65.1 (DAAE Horto)		
	Flooding	Visconde do Rio Claro Avenue		36 (Santana)		
	Flooding	Felicio Castellano Avenue, near the Arco-iris neighborhoods		63 (Brastemp)		
	Flooding	Downtown streets		36 (Santana)		
	Flooding	Jardim Sao Paulo		36 (Santana)		
March						
Date	**Type of Occurrence**	**Location**	**Precipitation (mm)**	**Rainfall station near the location (mm)**	**Obs:**	**Type of Atmospheric System**
16	Flooding	Tancredo Neves Avenue	24	31 (Santana)	7pm to 24h00m	Convective rainfall
	Flooding	Avenue		31 (Santana)		
		Tancredo Neves				
April						
Date	**Type of Occurrence**	**Location**	**Precipitation (mm)**	**Rainfall station near the location (mm)**	**Obs:**	**Atmospheric System Type**
10	Flooding	Av. 29 bridge - Vila Paulista	23,1	3.8 (Gessy Ineos)	5pm to 18h00m	Convective rainfall
	Flooding	Street 3 - Center		25 (Santana)		
	Flooding	6th Street - Center		25 (Santana)		
	Flooding	Visconde do Rio Claro Avenue		25 (Santana)		
October						
Date	**Type of Occurrence**	**Location**	**Precipitation (mm)**	**Rainfall station near the location (mm)**	**Obs:**	**Type of Atmospheric System**

1	Flooding (home)	Cherveson Garden	26,7	20 (Cavalry)	22h00m to 24h00m	Convective rainfall
29	Falling tree	Av. Felicio Castellano x Rua 19 - Arco-iris	43,1	49 (Brastemp)	6 p.m. to 24h00m	Convective rainfall
	Falling tree	Av. 25 and 27 - Bairro do Estàdio		66.1 (Civil Defense)		
	Fallen tree (2)/ Damage to telephone system	Av. Ulisses Guimaraes x Av. 10A - Vila Indaià		53.4 (DAAE Horto)		
	Damage to (unpaved) roads	Jardim Araucària		33 (Cavalry)		
	Mud	Av. 70 - Jardim Araucària		33 (Cavalry)		
December						
Date	**Type of Occurrence**	**Location**	**Precipitation (mm)**	**Rain station near the town**	**Obs:**	**Type of Atmospheric System**
				(mm)		
11	Flooding (home)	Av. 22 - n° 186 - Center	68,5	86 (Santana)	6 p.m. to 20h00m	Precipitation associated with the entry of a frontal system
	Flooding	Visconde do Rio Claro Avenue		86 (Santana)		
	Road damage	Jardim Inocoop		80.8 (Civil Defense)		
	Mud	Jardim Bom Sucesso		86 (Santana)		
16	Flooding	Av. Visconde do Rio Claro - Jardim Inocoop	16,1	29.3 (ETE Palmeiras)	11:00 a.m. to 12h00m	Convective rainfall
25	Collapse	Residential Jardim Novo - Jardim Novo II	32,3	60.8 (Agroceres)	02h00m to 06h00 and 6 p.m. to 19h00m	Precipitation associated with the entry of a frontal system

Source: Organized by LOPES, 2012

ANNEX E - Record of the amount of rainfall and its relationship with the location and types of occurrences and the atmospheric systems operating in 2009

January						
Date	**Type of Occurrence**	**Location**	**Precipitation (mm)**	**Rainfall station near the location (mm)**	**Obs:**	**Type of Atmospheric System**
3	Flooding	Jardim Inocoop	71,8	56.5 (ETE Palmeiras)	8pm to 22h30m	Precipitation associated with the entry of a frontal system
	Flooding	Avenida 6 x Rua 7 - Santa Cruz		57 (Santana)		
14	Flooding	Santa Casa (PSMI) and Day Hospital - Jardim Cherveson	28,4	51 (Cavalry)	1 p.m. to 16h00m	Convective rainfall
February						
Date	**Type of Occurrence**	**Location**	**Precipitation (mm)**	**Rainfall station near the location (mm)**	**Obs:**	**Type of Atmospheric System**
	Flooding	Jardim Inocoop		52.2 (ETE Palmeiras)		
	Flooding	Av. 19 x Street 9 - Center		46.5 (Santana)		
	Falling tree	Rua 19 x Av. 13 e 15; n°450 - Consolaçâo		75.5 (Civil Defense)		
	Falling tree	Rua 3 x Av. 37 e 39; n° 710 - Cidade Jardim		75.5 (Civil Defense)		
	Falling tree	Rua 8 x Av. 54 - Alto do Santana		46.5 (Santana)		
8	Collapse	Jardim Inocoop	24,7	52.2 (ETE Palmeiras)	09h00 to 10h00m	Convective rainfall
9	Flooding	Avenue Castelo Branco - Jardim Sâo Paulo	25,4	44 (Santana)	8 p.m. to 21h30m	Convective rainfall
12	Flooding	Street 2, n°	44,8	44 (Santana)	23h00m	Precipitation
		1740 - Center			at 24h00m	associated with the SACZ
March						
Date	**Type of Occurrence**	**Location**	**Precipitation (mm)**	**Rainfall station near the location (mm)**	**Obs:**	**Atmospheric System Type**
9	Flooding	Jardim Sâo Paulo parallel to Av. Castelo Branco	47	86.4 (Claretianas)	6pm to 21h30m	Convective rainfall
	Flooding	Av. Tancredo Neves - Jardim Inocoop		78.1 (ETE Palmeiras)		
12	Problems with the sewage system (home)	11th Street - Bairro Santa Cruz	28	37 (Santana)	23h00m to 24h00m	Precipitation associated with the SACZ
September						
Date	**Type of Occurrence**	**Location**	**Precipitation (mm)**	**Rainfall station near the location**	**Obs:**	**Atmospheric System Type**

				(mm)		
8	Flooding	Visconde do Rio Claro Avenue	62,2	31 (Santana)	14h00 to 15h00m	Precipitation associated with the induced trough
	Falling tree	Street 6 - Alto do Santana		31 (Santana)		
	Falling tree	Avenue 1 - Center		31 (Santana)		
December						
Date	**Type of Occurrence**	**Location**	**Precipitation (mm)**	**Rainfall station near the location (mm)**	**Obs:**	**Atmospheric System Type**
18	Falling tree	Pontilhao da Av. 7 x Rua 1 - Centro	24,4	26.5 (Santana)	16h00m to 21h00m	Convective rainfall
29	Flooding	Av. Tancredo Neves - Jardim Inocoop	119,7	74.7 (ETE Palmeiras)	24h00 to 06h00m	Convective rainfall
	Flooding	Castelo Branco Avenue		123.5 (Claretians)		
	Damage to buildings (cracks in 10 homes)	Av. 32 x Rua 6 e 7 - Santana		133 (Santana)		
30	Flooding	Tancredo Neves Avenue	33,7	0.8 (Claretians)	04h00m to 06h00m	Convective rainfall
	Flooding	Jardim Inocoop		28.9 (ETE Palmeiras)		

Source: Organized by LOPES, 2012

ANNEX F - Record of the amount of rainfall and its relationship with the location and types of occurrences and the atmospheric systems active in 2010

January						
Date	**Type of Occurrence**	**Location**	**Precipitation (mm)**	**Rainfall station near locality (mm)**	**Obs:**	**Type of Atmospheric System**
	Flooding	Basic Health Unit building - Vila Cristina		106 (Brastemp)		
	Flooding	Jardim Inocoop		86.4 (ETE Palmeiras)		
	Flooding	Pontilhao da Av. 29 x Rua 0 - Jardim Donângela		89.7 (Civil Defense)		
	Flooding	Pontilhao da Av. 7 x Rua 1 - Centro		119.9 (DAAE Horto)		
	Flooding	Av. 16 x intersection of Av. Castelo Branco - Parque Sao Paulo		100.4 (Claretianas)		
	Flooding	17th Street x Av. 38 - Jardim Bela View		114 (Santana)		
	Flooding	Avenue 50 x Street 3 and 1 - Jardim Portugal		114 (Santana)		
	Flooding	Rua 20 x Av. José Felicio Castellano - Arco-iris		106 (Brastemp)		
	Flooding	Avenue pontoon Tancredo Neves		100.4 (Claretianas)		
8	Flooding	Pointer Jardim Primavera	106,9	114 (Santana)	6 p.m. to 7 p.m. m	Convective rainfall
	Flooding (Cachoeirinha stream overflowed)	Black Mother		166 (Florence)		
	Flooding (home)	M-2 Street x Avenues M-5 and M-7 - Vila Martins		114 (Santana)		
	Flooding (home)	Avenida Visconde do Rio Claro x Rua 5 - Centro		114 (Santana)		
	Flooding (home)	Av. M-27 x Rua 6 - Jardim Hipódromo		106 (Brastemp)		
	Flooding (home)	Rua 4 x Av. 18 e 20 - Centro		114 (Santana)		
	Flooding (home)	Rua 14, n° 241 - Santa Maria		114 (Santana)		
	Flooding (home)	Av. 48, n° 1125 - Vila Nova		106 (Brastemp)		

	Flooding (residence)/ Collapse	Vila Nova		106 (Brastemp)		
	Flooding/ Road damage	Avenida Brasil		106 (Brastemp)		
	Road damage	Avenue 54 - Jardim Santa Eliza		114 (Santana)		
	Road damage	Av. 11 (back of the old Telesp Employees Club) - Jardim Esmeralda		86.4 (ETE Palmeiras)		
	Road damage	Av. 26 x Av. Visconde do Rio Claro		114 (Santana)		
	Sliding	Street 6 x Av. M-23 - Chàcara Boa Vista		106 (Brastemp)		
19	Flooding	Avenue 50A x Streets 7A and 8A - Vila Alema	48,5	45 (Brastemp)	16h00 m to 18h00 m	Convective rainfall
February						
Date	**Type of Occurrence**	**Location**	**Precipitation (mm)**	**Rainfall station near Iocality (mm)**	**Obs:**	**Atmospheric System Type o**
	Flooding/death	Tancredo Neves Avenue x SP 310 access loop		173.8 (Claretians)		
	Flooding	Avenida 1 x Tancredo Neves - Jardim Inocoop		55 (ETE Palmeiras)		
	Flooding	Street 3 - New Jd. Wenzel		173.8 (Claretians)		
	Flooding	Jardim Bom Sucesso		173.8 (Claretians)		
	Flooding	Tancredo Neves Avenue - Jardim Inocoop		55 (ETE Palmeiras)		
	Flooding	Visconde do Rio Claro Avenue		42 (Santana)		
	Flooding	Access loop to Tancredo Neves Avenue		173.8 (Claretians)		
	Flooding	Conduct Garden		33.2 (Gessy Ineos)		
25	Flooding	Pontilhao da Av. 7 x Rua 1 - Centro	61	87.4 (DAAE Horto)	21h00 m to 24h00 m	Precipitation associated with the entry of a frontal system
	Road damage	14th Street - Jardim Santa Maria		37 (Cavalry)		
	Road damage	Av. 54 - Jardim Santa Eliza		42 (Santana)		
	Road damage	Rua 8 x Av. Visconde do Rio		42 (Santana)		

		Claro				
	Damage to buildings (falling porches)	Av. 6 x Av. Visconde do Rio Claro		42 (Santana)		
	Problems with the sewage system	Rua 11 - Res. das Flores		37 (Cavalry)		

March

Date	**Type of Occurrence**	**Location**	**Precipitation (mm)**	**Rainfall station near Iocality (mm)**	**Obs:**	**Atmospheric System Type o**
15	Flooding	Avenida 7 x Rua 1; pontoon - Center	45,8	37 (Santana)	03h00 m to 07h00 m	Convective rainfall

Decem bro

Date	**Type of Occurrence**	**Location**	**Precipitation (mm)**	**Rainfall station near the location (mm)**	**Obs:**	**Atmospheric System Type o**
5	Flooding	Avenida 16 - Avenida Castelo Branco - Jardim Sao Paulo	48,3	21 (Claretians)	6 p.m. to 7 p.m. and 23h00 m to 24h00 m	Convective rainfall
	Flooding	Visconde do Rio Claro Avenue		56 (Santana)		
	Flooding	Bus Terminal area on Av. Tancredo Neves		21 (Claretians)		
	Falling tree	Avenue 3 - Vila Industrial		49 (Brastemp)		
	Road damage	Tancredo Neves Avenue		21 (Claretians)		
	Problems with the sewage system / Problems with the water supply	Jardim Inocoop		45.3 (ETE Palmeiras)		
	Damage to erosion works	Jardim Esmeralda		45.3 (ETE Palmeiras)		
12	Mud	New Garden	29,6	15.4 (ETE Palmeiras)	17h00 m to 21h00 m	Convective rainfall
	Mud	Guanabara Garden		15.4 (ETE Palmeiras)		
	Falling tree	Rua 8 x Avenida da Saudade - Center		41.5 (Santana)		
	Falling tree	Rua 6 x Av. 40 and 42 - Alto do Santana		41.5 (Santana)		
13	Flooding	Rua 21 - Jd. Nova Rio Claro	18,4	24.6 (ETE Palmeiras)	04h00 m to 12h00 m and 13h00 m to 15h00 m	Precipitation associated with the entry of a frontal system
	Flooding	Jardim América		58.3 (Brastemp)		
	Falling tree	Rua 11A x Av. 24 e 26A - Vila Alema		41.5 (Santana)		
	Falling tree	Av. 44A x Rua		58.3 (Brastemp)		

		5A - Jardim Ipê				
	Falling tree	Rua 4-J A and 5-J A x Av. 54A and 56A - Jardim América		58.3 (Brastemp)		
	Falling tree	Rua 6 x Av. 25 - Cidade Jardim		11.1 (Civil Defense)		
	Fallen tree (affected the power grid)	Av. José Felicio Castelano -		58.3 (Brastemp)		
		Vila Cristina				
	Sliding	Jardim Inocoop		24.6 (ETE Palmeiras)		
	Problems with the sewage system	Rua 9 X Av. 64 and 66 - Jardim América		58.3 (Brastemp)		
21	Damage to roads (sidewalks)	Street 3 - Jardim Portugal	8,8	10 (Cavalry)	02h00 m to 03h00 m	Convective rainfall
	Flooding	Stretches of Av. 16 - Jardim Sao Paulo		52.4 (Claretianas)		
	Flooding	Av. Visconde do Rio Claro x Rua 14 - Vila do Ràdio		52.4 (Claretianas)		
	Flooding (the Córrego da Servidao overflowed)	Jardim Inocoop		46.9 (ETE Palmeiras)		
	Slipping (construction site)	Av. Visconde do Rio Claro x Rua 14 - Vila do Ràdio		52.4 (Claretianas)		
26	Road damage	Av. Tancredo Neves	41,4	52.4 (Claretianas)	14h00 m to 15h00 m	Convective rainfall

Source: Organized by LOPES, 2012

MIX
Papier aus verantwortungsvollen Quellen
Paper from responsible sources
FSC® C105338

Printed by Books on Demand GmbH, Norderstedt / Germany